Eggs for Breakfast

By Annette Smith
Photographs by Lindsay Edwards

Here is my family.

We are going
to have breakfast.

This mat is for Mom.

This mat is for Dad.

This mat is for my brother.

And this mat is for me.

The mats are on the table.

Mom's plate goes here.

Dad's plate goes here.

My brother's plate goes here.

And my plate goes here.

Here is a spoon
for Mom.

Here is a spoon
for Dad.

Here is a spoon
for my brother.

And here is a spoon
for me.

We are going to have eggs for breakfast.

Here are the egg cups

for the eggs.

13

Here is an egg
for Mom.

Here is an egg
for Dad.

Here is an egg
for my brother.

And here is my egg.

We like eggs for breakfast.